FARIDA ALLAL
Khouloud KOURIMI
NIHAL KHATOUI

Antibacterial activity of Celecoxib

FARIDA ALLAL
Khouloud KOURIMI
NIHAL KHATOUI

Antibacterial activity of Celecoxib

ScienciaScripts

Imprint
Any brand names and product names mentioned in this book are subject to trademark, brand or patent protection and are trademarks or registered trademarks of their respective holders. The use of brand names, product names, common names, trade names, product descriptions etc. even without a particular marking in this work is in no way to be construed to mean that such names may be regarded as unrestricted in respect of trademark and brand protection legislation and could thus be used by anyone.

Cover image: www.ingimage.com

This book is a translation from the original published under ISBN 978-620-6-71017-2.

Publisher:
Sciencia Scripts
is a trademark of
Dodo Books Indian Ocean Ltd. and OmniScriptum S.R.L publishing group

120 High Road, East Finchley, London, N2 9ED, United Kingdom
Str. Armeneasca 28/1, office 1, Chisinau MD-2012, Republic of Moldova, Europe
Managing Directors: Ieva Konstantinova, Victoria Ursu
info@omniscriptum.com

Printed at: see last page
ISBN: 978-620-8-57812-1

TABLE OF CONTENTS

INTRODUCTION

Excessive exposure to antibiotics is contributing to the spread of bacterial resistance, representing a major public health problem. Antibiotic resistance has been on the rise for several decades, making it difficult to treat patients. It leads to an increase in the length of treatment and in the morbidity associated with infections.

Developing new antibiotic molecules can be long, tedious and costly. Faced with new threats, scientists have turned to innovative therapeutic strategies such as passive immunisation by injecting antibodies against bacterial virulence factors, antimicrobial peptides, bacteriophages and molecular repositioning, which involves using a drug that has already been approved for a new therapeutic indication to treat another disease (**GAJDÁCS, 2020**) (**Moshawih et al ,2024**).

Several advantages are associated with this strategy, such as reducing the high costs associated with developing new molecules, including research and development costs, as well as the costs of initial clinical trials, speed of development and prior knowledge of pharmacological and toxic properties (**Schlesinger et al, 2009**) (**Thangamani, 2015**)

This is the background to this dissertation, main aim of which is to use a non-antibacterial drug to combat bacterial resistance to antibiotics. The drug in question is Celecoxib (CXB), a non-steroidal anti-inflammatory agent in the class of selective COX-2 inhibitors, used to treat a variety of inflammatory and painful conditions, including osteoarthritis, rheumatoid arthritis, ankylosing spondylitis and acute pain. It can also be used to relieve pain associated with dysmenorrhoea. According to the Biopharmaceutical Classification System (BCS), CXB belongs to class II, and is characterised by low solubility and high permeability. It can therefore benefit from advanced formulation techniques to improve its solubility and bioavailability, such as the formation of co-crystals, its inclusion in polymeric matrices, the formation of salts and solvates, the formation of nanoparticles, the formation of micro and nano emulsions and the development of eutectic mixtures (**Hyun et al, 2018**) (**Yang et al, 2022**). These approaches aim to overcome solubility limitations to maximise drug absorption and therapeutic efficacy. In this study, repositioning strategies and the formation of eutectic mixtures by binary combination of CXB with different coformers were chosen to combat multi-resistant bacteria and to increase the solubility, bioavailability and therefore the therapeutic effect of CXB. The choice of coformer is based on its chemical structure, the presence of functional groups, proton donors and acceptors, capable of establishing non-covalent intermolecular interactions, such as hydrogen bonding, Van Der Waals forces and r-r interactions (M Singh et al,2023). The different coformers chosen are: salicylic acid (A.SAL), maleic acid (A.MAL), nicotinamide (NIC), succinic acid (A.SUC) and cinnamic acid (A.CIN).The aim of this work is to explore these innovative strategies for combating bacterial antibiotic resistance and improving solubility. This dissertation is divided into three chapters: :The first chapter presents a review of the literature, explaining the concept of repositioning drug molecules and the approach to forming eutectic mixtures to improve the physicochemical properties of drugs. The second chapter describes the equipment used, the experimental protocols for synthesis and characterisation using various techniques, namely differential scanning calorimetry (DSC), X-ray powder diffraction (XRD), scanning electron microscopy (SEM) and Fourier Transform Infrared (FTIR), the evaluation of antibacterial activity using three bacterial strains and the determination of solubility. The third chapter deals with the results of the synthesis, characterisation, evaluation of the in vitro antibacterial activity and determination of the solubility of the eutectic mixtures produced.The results obtained and any future prospects will be summarised in the general conclusion.

SUMMARY BIBLIOGRAPHY

I.1. Bacteria

I.1. Bacteria

Bacteria are single-celled micro-organisms capable of reproducing independently. They range in size from 1 to 10 µm and weigh from 10 to 12 grams. Bacteria do not have a nucleus bounded by a membrane to isolate their genetic material. Genetic information is supported deoxyribonucleic acid (DNA), assembled in the form of one or more chromosomes whose structure is a double helix of circular, coiled DNA, supercoiled in the cytoplasm thanks to the action of topoisomerases (4 in number in bacteria) **(M.Bouskraoui et al.,2017)** . As a prokaryote, the structure of the cell is simple. The interior volume, called the cytoplasm, is bounded by the plasma membrane, which controls the flow of energy into and out of the bacterium and serves as a support for certain enzymes. The cytoplasm is particularly rich in soluble RNA (messenger RNA and transfer RNA) and, above all, particulate or ribosomal RNA. Ribosomes are the target of action of many antibiotics such as aminoglycosides, phenicols, cyclins and macrolides. The cytoplasmic membrane or inner membrane is the outer boundary of the cytoplasm. It is made up of a double layer of phospholipid units (35%) and associated proteins (65%). Some of these proteins play a role in peptidoglycan synthesis and are known as penicillin-binding proteins (PBP) because they are also the target of action of beta-lactams, the family of antibiotics to which penicillin belongs. In addition to the chromosome, the bacterium may contain small genetic elements (DNA) (0.5 to 5% of the bacterial chromosome) that are extra-chromosomal. These elements, called plasmids, are not essential to the life of the bacterium under normal growth conditions. They replicate independently and generally more rapidly than the bacterial chromosome **(M.Bouskraoui et al.,2017)**.

.

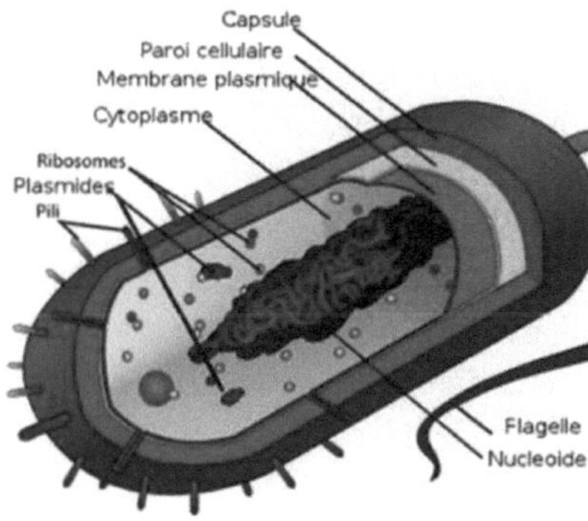

Figure I.1. Diagram a bacterium (Wikipedia)

Beyond the cytoplasmic or plasma membrane, bacteria have a wall made up of peptidoglycan. This ensures the mechanical cohesion of the cell. It remains permeable to Gram dye and the corresponding bacteria are known as Gram-positive. Other bacteria have a simpler peptidoglycan wall and, above all, an impermeable outer membrane: these are Gram-negative bacteria.

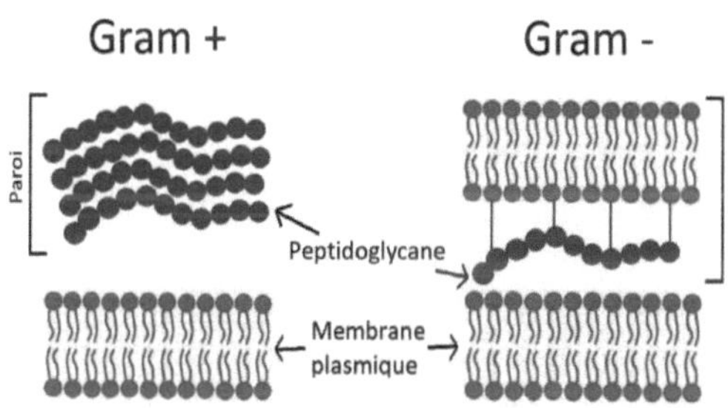

Figure I.2. Bacterial wall of gram-positive and gram-negative bacteria (Wikipedia)

I.2. Bacterial resistance

Bacterial infections remain a major cause of morbidity and mortality, and bacterial chemotherapy is becoming increasingly limited, despite the large number of antibiotics that have been discovered. Antibiotics have played an important role in reducing mortality from infectious diseases, but their massive and repeated use has led to the emergence of multi-resistant strains. Antibiotic resistance is one of the major threats to human health today, according to the

World Health Organisation (WHO). Bacterial resistance has been steadily increasing for several decades, complicating patient treatment, prolonging its duration, increasing the morbidity associated with infections and potentially jeopardising vital prognosis **(K. Amanda et al., 2024; H. Lode et al., 2009).**
Growing concern about multi-resistant bacterial strains calls for the development of additional bactericidal means. One of the challenges facing research today is to apply innovative new strategies such antibody , antimicrobial peptides, bacteriophages and molecular repositioning. However, the use of non-antibacterial agents such as antihistamines, antidepressants, antihypertensives, antipsychotics and non-steroidal anti-inflammatory drugs is making it possible tofind of solutions innovative solutions for fight against these strains multi-resistant bacterial strains. A therapeutic repositioning strategy was therefore chosen in this study, involving the use a non-steroidal anti-inflammatory drug, celecoxib, to combat different bacterial strains.

I.3. Repositioning of active ingredients

Drug repositioning or drug repurposing is a promising approach. This alternative strategy involves exploring existing drugs and applying them to diseases other than those for which they were initially developed. It offers a great opportunity rapidly identify a new drug candidate at a lower cost than the discovery and development of new drugs. Drug repositioning is a highly efficient process, without the need for a large investment. risk compared to new drug development, as existing knowledge and data on drugs, such as pharmacokinetics, formulation, pharmacology, toxicity and manufacturing data, are available. Drug repositioning also reduces the number of steps required for Food and Drug Administration (FDA) approval, which could potentially speed up a drug's time to market **(Kushwaha et al., 2023; F.Ali et al., 2024)**. There is little literature on the use of existing drugs to treat other diseases. **(S.Thangamani et al., 2015)** have shown that celecoxib exhibits broad-spectrum antimicrobial activity against Gram-positive pathogens of various genera, including Staphylococcus, Streptococcus, Listeria, Bacillus and Mycobacterium.
The results of research into liver diseases carried out by **(F.Ali et al., 2024)** highlight the importance of repositioning drugs as a tool to address the global burden of liver diseases, including hepatitis, cirrhosis of the liver, Wilson's disease, liver cancer and fatty liver.

I.4. Non-steroidal anti-inflammatory drugs (NSAIDs)

Non-steroidal anti-inflammatory drugs (NSAIDs) are a widely used therapeutic class, thanks their antipyretic, analgesic and anti-inflammatory properties. They are used in a wide range of indications, including osteomuscular pain, rheumatological conditions , in traumatologie and in in colic nephritis **(F. Pillon.,2014)**. These are symptomatic drugs capable of opposing the inflammatory process, whatever the cause: mechanical, chemical, infectious or immunological. They belong to several chemical families and are characterised by the absence of a steroidal chemical structure **(F. Pillon.,2014)**. They constitute a chemically diverse group of molecules: such as salicylic acid derivatives (Aspirin) propionic acid derivatives (Ibuprofen and Naproxen), and others (Diclofenac, Indomethacin, Nabumetone), COX-2 selectives (Etoricoxib, Celecoxib, Lumiracoxib), but they are all capable inhibiting COX cyclooxygenase and the resulting attenuation of prostaglandin synthesis **(H.Blain, 2000)** They are effective and prescribed for the treatment of a variety of inflammatory and painful conditions **(D.Muster..., 2005)**.

I.5. Mode of action of NSAIDs

Although their chemical composition is different, non-steroidal anti-inflammatory drugs (NSAIDs) have a common mode of action. They inhibit cyclooxygenase by penetrating the channels that make up these enzymes, blocking them by binding via hydrogen bridges to an arginine in the lower half of the channel. This reversibly inhibits these enzymes by blocking access to arachidonic acid, thereby reducing tissue synthesis of prostaglandins (PGs) (Figure I.1). The pharmacological effects of all the NSAIDs shown in the diagram below result from inhibition of prostaglandin synthesis, which leads to opposition to the actions of these substances **(N.Levion., 2002 ; H.Blain., 2000)**

.

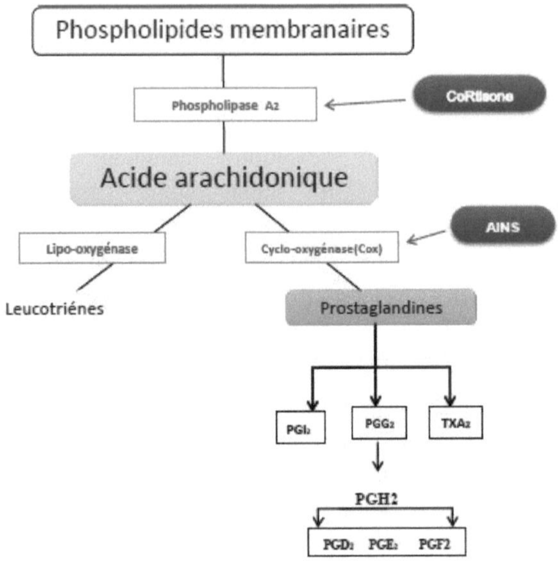

Diagram I.1. Metabolism of arachidonic acid and mode action of NSAIDs

I.5. Pharmacological effects of NSAIDs

NSAIDs are commonly used because of their beneficial pharmacological properties. However, their use is also associated with a number of adverse effects. Table 1 describes the pharmacological consequences of NSAIDs

Table I.1. Effects of prostaglandins and NSAIDs

Targets	Effect of prostaglandins	Effect of NSAIDs
SNC	Raises the threshold at which hypothermic regulations are triggered	Antipyretic
Pain	+	Analgesic
Platelets (COX-1 alone)	Via thromboxane progrégant	Anti-aggregant
Stomach	Inhibition of acid secretion + increase in mucus secretion	Ulceritis
Kidney	Regulation of renal blood flow in hypovolaemia; vasodilation in the failing kidney	Decreased renal blood flow; renal failure
Bronchial tubes	Expansion	Bronchoconstriction
Inflammation	Pain, redness, heat	Anti-inflammatory
Uterus	Contraction	Delays / painful contractions during menstruation or childbirth

I.6. Non-steroidal anti-inflammatory drug, selective COX-2 inhibitor: Celecoxib

The celecoxib or (4-[5-(4-méthylphényl) -3-(trifluoromethyl) -pyrazol-1-yl]) - benzenesulfonamide), is a non-steroidal anti-inflammatory agent used to treat rheumatoid arthritis, osteoarthritis, acute and chronic pain and inflammation **(M. Srinivasu, 2002)**. In addition, CXB is recognised for its anti-carcinogenic potential, notably for colon tumours, breast cancer and skin cancer. Its anti-inflammatory, analgesic, antipyretic and anticancer effects stem from its selective inhibition of cyclooxygenase-2 (COX-2), an essential enzyme involved prostaglandin biosynthesis **(S.Dadiboyena et al.,2012; L.Tive.,2000; I.Tołoczko et al.,2019)**. Celecoxib has also been used for its antimicrobial properties in the fight against resistant bacteria. For example, **(Salunke et al., 2006)** showed the efficacy of celecoxib and its derivatives in the fight against tuberculosis caused by Mycobacterium tuberculosis. **(H.Chiu et al.,2012)** evaluated the antimicrobial efficacy of celecoxib in the in-vitro eradication of Staphylococcus aureus, a Methicillin-resistant bacterium. The results showed the bactericidal potential of celecoxib against Staphylococcus aureus. This significant antibacterial action prompted researchers to use other agents based on structural modification of CXB. Chiu and colleagues **(H.Chiu et al., 2009)** tested celecoxib on the bacterial strain "Francecisella tularensis", a resistant gram-negative bacterium responsible for tularemia, an infection with a mortality rate of 60%. The results obtained demonstrated the effectiveness of celecoxib on this bacterium through intracellular proliferation. **(SA. Alshawwa et al., 2022)** tested CXB-loaded nanoparticles as an in vivo therapeutic approach to combat infection caused by Staphylococcus aureus. This innovative technology represents a significant advance in the treatment of antibiotic-resistant bacterial infections. It is based on the encapsulation of CXB and the formation of nanoparticles capable of penetrating infected tissue for targeted, controlled delivery. Because of their unique structure and large specific surface area, these nanoparticles can improve the stability and bioavailability of CXB, while minimising potential side effects. In previous work, schiff bases containing the pyrazole nucleus have been tested. for their antibacterial activity in vitro by the agar diffusion method against two Gram-positive bacteria (Bacillus subtilis (ATCC no. 23857) and Staphylococcus aureus (ATCC-29213)) and two Gram-negative bacteria (Escherichia coli (ATCC-25922) and P. aeruginosa (ATCC no. 25619)). Ciprofloxacin was used as a positive control. The results obtained showed good antibacterial activity, with MICs of 0.78 μg/ml and 1.562 μg/ml, respectively, against the Gram-positive bacterial strain S. aureus. Another study by **(S.Thangamani et al., 2015)** showed the efficacy of celecoxib on gram-positive and gram-negative bacterial strains. These included Staphylococcus, Streptococcus, Listeria, Bacillus, Mycobacterium, Acinetobacter and Pseudomonas, by inhibiting their RNA, DNA and protein synthesis. The in vitro antibacterial activity of a number of non-steroidal NSAIDs, namely MeloxicamCelecoxib, Valdicoxib and Nimesulide, was evaluated against 20 bacterial strains taken from human skin and urine. The antibacterial effect, determined by measuring zone of inhibition diameters and minimum inhibitory concentration, showed that all the bacteria tested were sensitive to all the Coxibs used, with the exception of Pseudomonas aeruginosa, Klebsiella pneumoniae and Staphylococcus aureus, which were resistant to Meloxicam, Nimesulide and Valdicoxib respectively **(H. Alkuraishy, 2013).**

I.7. Mechanism of action of celecoxib

Celecoxib acts by selectively inhibiting the cyclooxygenase-2 (COX-2) enzyme, responsible for the production of inflammatory prostaglandins (Diagram I.2).
Prostaglandins are chemical compounds that the body naturally produces in

response to inflammation or injury. They are essential for managing pain, inflammation and fever. By specifically inhibiting COX-2 activity, celecoxib reduces the production of inflammatory prostaglandins without having any significant effect on COX-1, another form of cyclooxygenase enzyme responsible for the production of prostaglandins involved in protecting the gastric mucosa and regulating platelet function **(RN.Saha et al.,2002)**.
In this way, by reducing the secretion of inflammatory prostaglandins, CXB helps to alleviate inflammation, relieve pain and reduce fever in patients suffering from inflammation, such as osteoarthritis, rheumatoid arthritis and ankylosing spondylitis. The specific inhibition of COX-2 by CXB is a viable approach to reducing inflammation. gastrointestinal toxicity previously associated with traditional NSAIDs, while providing the efficacy associated with high doses of these drugs. **(M.Cervello et al.,2011)**.

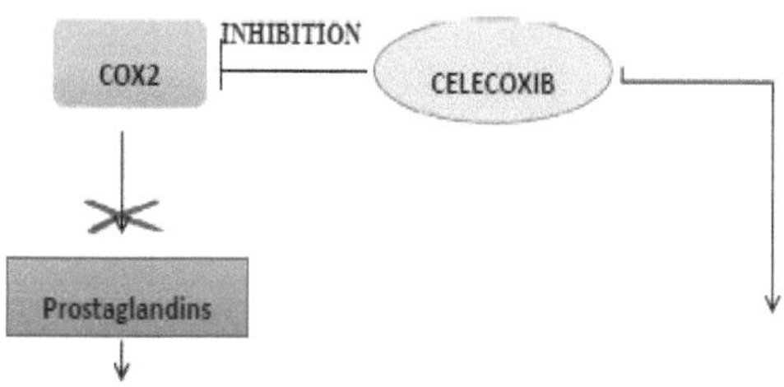

*Increased inflammation *Decreased production of prostaglandins

*Increased apoptosis *Decreased pain and inflammation

*Increased vascular permeability * Reduces pathological retinal angiogenesis

*Leukocyte migration *Reduces vascular leakage

*Intraocular inflammation

Diagram I.2. Mechanism of action of celecoxib.

Celecoxib (CXB) belongs to category II of the biopharmaceutical classification system (BCS), due to its low solubility in water, combined with high permeability. This limited solubility poses challenges for drug dissolution, affects oral bioavailability and may lead to therapeutic deficiencies in drug candidates. To address these challenges, various methodologies have been explored to improve the solubility, dissolution rate and oral bioavailability of poorly water-soluble drugs, including amorphous solid dispersions, co-amorphous systems, micro- and nanoparticles, salt formation, nanocrystals, cocrystals, inclusion and eutectic mixtures which are the focus of this study.

I.8. Mixture eutectic

In the pharmaceutical industry, a eutectic mixture can be prepared by combining a drug and an inert carrier (usually a highly hydrophilic compound) or by combining two drugs with different solubilities. This combination is based on the establishment of intermolecular interactions known as weak adhesive interactions, such as as shown schematically in **Figure I.3 (G.C. Bazzo et al., 2020)**.

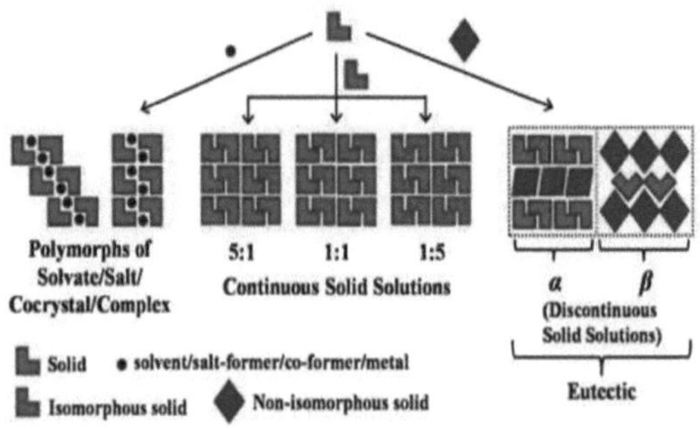

Figure I.3: Schematic representation of the structural structure of a eutectic mixture (G.C. Bazzo et al., 2020).

The potential of such a mixture lies in increasing the solubility, dissolution rate or bioavailability of orally administered drugs. **(R.A. Alshaikh et al., 2019; G. C. Bazzo et al., 2019; R.A. Riekes et al., 2016)**.

Eutectic mixtures have continued to attract significant interest as a drug delivery system for improving the oral bioavailability of poorly water-soluble drugs, for the following reasons:

- Preparation methods are cheaper and easier to produce;
- Eutectics are not considered to be new chemical entities or new crystalline forms and therefore do not require clinical trials.
- In eutectic systems, both components exist in crystalline form, which are very stable compared to amorphous materials **(Araya-Sibaja et al., 2019).**

Several studies have focused on the combination of a class II or IV active ingredient with an inert carrier or another drug, in order to increase its oral bioavailability. **(S.M. Hyun et al., 2019)** synthesised and characterised two eutectic mixtures, obtained by the binary combination celecoxib/adipic acid and celecoxib/saccharin. The results offered a new formulation strategy to improve the solubility and oral bioavailability of celecoxib.

S. Emami et al, 2018: J. Haneef et al., 2017; S.M. S. Hyun et al., 2019) have developed eutectic mixtures using various organic acids, such as adipic acid, succinic acid, ascorbic acidfumaric acid, gallic acidmalonic acid, nicotinamide and some amino acids **(K. Chahda et al, 2017; R. Chahda et al., 2017; Park et al., 2020; Haneef et al., 2018).**

I.9. Solubility

The solubility of a product consists of its dissolution in a given solvent. The solubility of a solute by a given solvent only occurs if the solvent-solvent bonds (in the solution) and solute-solute bonds (in the crystal) can be replaced by solvent-solute bonds. Depending on the type of solvent, there are 4 types of interaction with the solute: ion-dipole, dipole-dipole, hydrogen bonding and complex formation.

I.10. Co-formers

Coformers are compounds which, by interacting with the active ingredient, can improve its physico-chemical characteristics, in particular its solubility. The choice of coformer is crucial to the formation of an effective eutectic mixture.
Potential coformers must :

□ Be compatible with celecoxib;

□ Adequate solubility and bioavailability;

□ Do not interfere with the pharmacological activity of celecoxib.

PART EXPERIMENTAL

This study consists of evaluating the antibacterial activity and solubility of eutectic mixtures obtained by combining a non-steroidal anti-inflammatory drug, celecoxib (CXB), and five organic molecules acting as coformers, belonging to different chemical families and possessing functional groups capable of establishing different types of intermolecular interactions, namely: dipole-dipole interactions, Van-Der-Waals forces, specific hydrogen bonding interactions and n- and o- interactions. n. The binary mixtures prepared in a 1:1 stoichiometric ratio are :

- Celecoxib-maleic acid
- Celecoxib-Cinnamic acid
- Celecoxib-Salicylic acid
- Celecoxib-nicotinamide
- Celecoxib-Succinic acid

The synthesis and characterisation of the above-mentioned mixtures were carried out at the Physical Chemistry of Materials Laboratory (LPCM) at Amar Telidji University and the Physical Chemistry Analysis Technical Platform (PTAPC. CRAPC Laghouat) respectively. Antibacterial activity was assessed at the Regional Veterinary Laboratory. A summary of the various experimental stages carried out is shown in the diagram below:

Preparation of eutectic mixtures

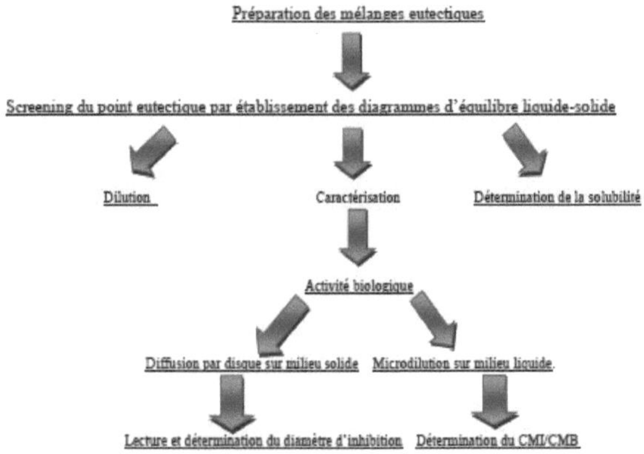

Diagram II.1: Summary diagram of the experimental procedure followed

II.1. Materials and methods

II. 1.1. Chemical products used

The chemicals used, which are summarised in **Table II.1**, were not further purified.

Table II.1: Chemicals used.

	Gross formula	Molecular weight (g/mol)	Purity(%)	Source
Celecoxib	C17H14F3N3 O2S	381.37	-	National Control Laboratory (LNC)
Nicotinamide	C6H6N2O	122,12	≥ 99	Sigma-Aldrich
Salicylic acid	C7H6O3	138.12	≥ 99.5	Sigma-Aldrich
Cinnamic acid	C9H8O2	148.15	≥ 99	Sigma-Aldrich
Succinic acid	C4H6O4	118.09	≥ 99.5	Sigma-Aldrich
Maleic acid	C4H4O4	116.07	≥ 99.5	Sigma-Aldrich

The physico-chemical properties of the products used in this work are shown on the **(Table II.2)**

Table II.2: Physico-chemical properties of the chemicals used.

Structure	Gross formula	Number of hydrogen bond donors	Number of acceptors hydrogen bonding	pKa	Solubility	Physical condition	Melting point
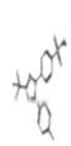 celecoxib					the		
					celecoxib		
					is little		
					soluble		
					in water.		
					Sa		
	C17H14F3N3 O2	2	5 à 6	11,0	solubility	powder	°

	S				in water	white	C
					is		
					of approximately		
					3à22		
					µg/ml.		
					However		
					it is more		
					soluble inorganic solvents such as chloroform and methanol		
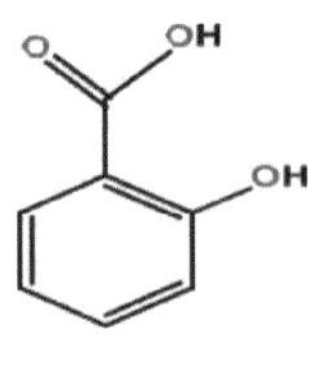					Soluble		
					in water		
					à		
					temperature	Powder	
	C7H6O3	2	2	3,0	e ambient.	Blanche	°C
Acide salicylique					Sa		
					solubility		
					in water		
					is		
					of approximately		
					2 g/L at		
					25°C.		
					However		
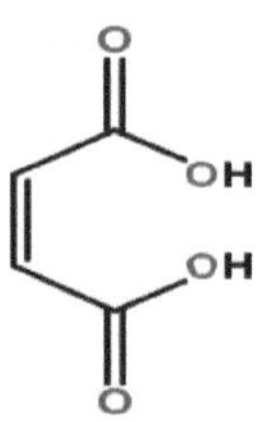					The acid		
					maleic		
					is soluble	crystals	
	C4H4O4	2	2	6,20	in water à a	colourless to	135°C
					concentrate	white	
					ion		
Acidemaléique					of approximately		
					553 g/L		
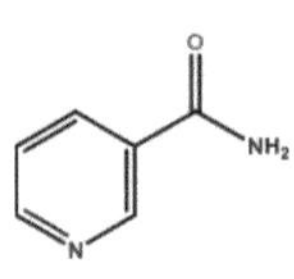					Very		
					soluble		
					in water		
	C6H6N2O	1	2	3.35	; 1 g soluble	Powder Blanc	°

nicotinamide						he	
					in 1		C
					ml of water		
					Soluble		
					in		
					butanol,		
					chlorofor		
					me		
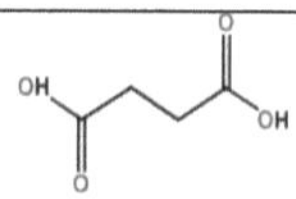					The acid		
					succinic		
					is soluble		
					in water	crysta ls	
Acide succinique	C4H6O 4	2	2	4,21	for good reason of approxi mately	colour less s to	185° C
					100 g/L.	white	
					However		
					his		
					solubility in other organic solvents may vary.		
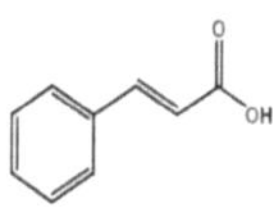					East		
					soluble		
					in water		
					à		
Acidecinnamique	C9H8O 2	2	2	4,44	temperat ure e	crysta ls white	134° C
					ambient.		
					Sa		
					solubility		
					in water		
					is		
					of approxi mately 2.1 g/L at		
					25°C.		
					However		
					it is more		
					soluble		
					in		
					solvents		
					organic		
					s.		

II.1.2. Microorganisms tested

The microorganisms chosen for this study are reference strains. They summarised in Table II.3.

Table II.3: General information on the bacterial strains used

Bacterial strains tested	Bacteriological character	Habitats	Pathogenicity	Standard strain identification number
Staphylococcus aureus	Gram+ Optional anaerobic	Man is the (M.Bouskraoui et al, 2017).	Superficial suppurative infections or deep: skin, soft tissue, muscle, bone. Toxinfection: due to the synthesis different toxins by certain strains. (M.Bouskraoui et al, 2017).	ATCC 25923
Escherichia coli	Gram - Facultative anaerobic	Normal host of the digestive tract. M.Bouskraoui et al, 2017).	Enterocolic infections, urinary tract infections, food poisoning, intra-abdominal infections, infections neonatal (meningitis). (M.Bouskraoui et al, 2017).	ATCC 25922
Klebsiella pneumoniae	Gram - Facultative anaerobic	Natural cavities, in particular the digestive tract and the upper airways. (M.Bouskraoui et al, 2017).	Nosocomial and community-acquired infections: bronchopulmonary and urinary tract infections. Purulent meningitis and sepsis... (M.Bouskraoui et al, 2017).	ATCC 70603

II.2. Methods experimental

II.2.1. Synthesis of eutectic mixtures

II.2.1.1. Liquid assisted grinding (LAG)

Liquid-assisted grinding involves adding a small volume of solvent to the active ingredient/coformer mixture to facilitate the formation of a eutectic mixture. The solvent in question acts as a catalyst, enabling the establishment of the interactions necessary for the formation of such mixtures **(A. Thayyil et al 2020)**. In order to prepare the various eutectic mixtures studied to assess their antibacterial activity and determine their solubility, we proceeded as follows: Using an Ohaus analytical balance accurate to $10^{(-4)}$ g, masses of celecoxib and the various coformers (inert carriers) were weighed in a 1:1 stoichiometric ratio **(Table II.4).** The weighed masses m_1 and m_2 were mixed and ground by the solvent-assisted liquid milling process, using a mortar and pestle for 30 min **(Figure II.1)**. Once synthesised, these equimolar mixtures are placed in an oven to allow evaporation of the solvent.

Figure II.1: LAG liquid-assisted grinding (Original, 2024).

Table II.4: Masses m_1 and m_2 of celecoxib and the various coformers.

	m1(g)	m2(g)
Celecoxib-maleic acid	0,35	0,11
Celecoxib-salicylic acid	0,42	0,15
Celecoxib-cinnamic acid	0,35	0,14
Celecoxib-nicotinamide	0,47	0,15
Celecoxib-succinic acid	0.35	0.11

II.3. Characterisation of synthesised eutectic mixtures

II.3.1 Differential Scanning

Differential Calorimetry (DSC) is a thermal analysis technique used to characterise changes in the state, phase or structure of a material. It is used to measure the heat flux exchanged by a sample and the oven relative to an inert reference, as a function of temperature and time, when they are subjected to the same heating or cooling programme (where T varies linearly with t). The temperature difference is converted into a heat flux difference in mW. This technique can be used to identify thermal accidents that a compound may undergo. These accidents may be reversible, such as melting or crystallisation, or irreversible, such as degradation or a chemical reaction caused by heating. To carry out a calorimetric analysis Differential Scanning Calorimetry (DSC), we used a Setaram 131 DSC apparatus (**Figure II.2**), fitted with aluminium crucibles of 30 μl volume and platinum thermocouples. The sample size was approximately 1 to 2 mg. The analysis was carried out in a nitrogen atmosphere at a heating rate of 2°C/min, over a temperature range from 20°C to 250°C. Temperatures were calibrated using indium, which has a melting temperature and enthalpy of 156.6°C and 28.5 J/g respectively.

Figure II.2: DSC used to measure heat flux (Original)

II.3.2. X-ray diffraction (XRD)

X-ray diffraction is one of the physicochemical analysis methods used to characterise crystal lattices. X-rays, like all electromagnetic waves, cause a displacement of the electron cloud relative to the nucleus in atoms; these induced oscillations result in the re-emission of electromagnetic waves of the same frequency. Depending on the direction in space, a large or small flow of X-ray photons takes place, and these variations in direction form the phenomenon of X-ray diffraction. Using Bragg's law, we can determine the lattice spacing For X-ray diffraction analysis, we used an EMPYREAN diffractometer (**Figure II. 3**). X-rays were emitted using Cu(Ka) radiation at room temperature over a range of 2 theta from 2 to 50°, with a step size of 0.01° and a scan speed of 2°/min.

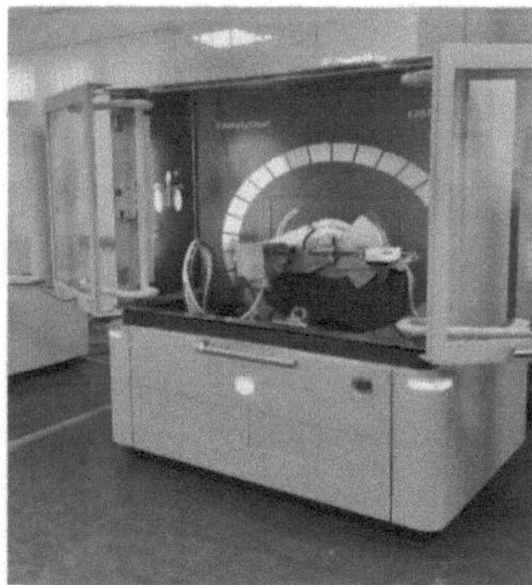

Figure II.3. Diffractometer type EMPYREAN (Original, 2024).

II.3.3. Scanning electron microscopy SEM

Scanning electron microscopy is a technique used to visualise the dispersive behaviour of a material. The principle involves scanning the surface of the sample with a very fine beam electrons that scans the surface point by point. Under impact of this accelerated electron beam, backscattered electrons and secondary electrons emitted by the sample are selectively collected by detectors, which transmit a signal to a cathode-ray screen, the scanning of which is exactly synchronised with the scanning of the object. This technique gives us information about the topography, morphology and composition of the sample. The equipment used is a Thermo Fischer SCIO2S environmental scanning electron microscope **(Figure II.4)**. Scanning electron microscopy analysis shows the shape and size of the particles obtained on a microscopic scale.

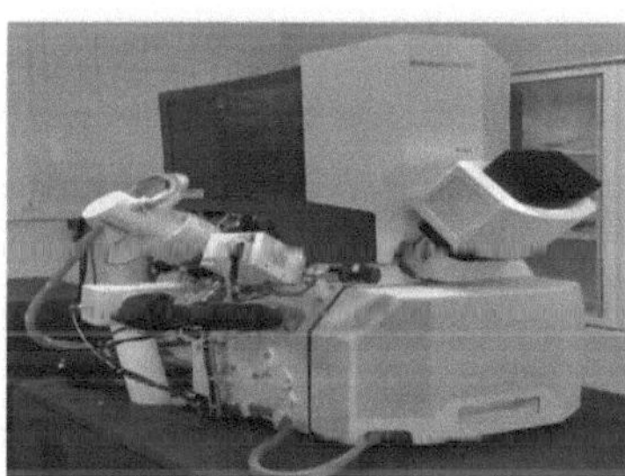

Figure II.4: Thermo Fischer SCIO2S environmental scanning electron microscope (Original, 2024).

II.3.4. Fourier Transform Infrared FTIR

II.3.4. Fourier Transform Infrared FTIR

Infrared spectroscopy is mainly used to study the interactions between molecules, by analysing the profile of the vibratory mode, i.e. the position, width and intensity of the spectral bands. The position of the peak or band not only indicates the presence of a particular group, but also gives a good idea of the environment affecting it. It is well known that intermolecular forces due to hydrogen bonding interactions cause a remarkable change in some of the vibrational modes, making it possible to study the interactions.

The formation of hydrogen bonds is of vital importance in many industrial processes, and plays a central role in biological processes at the molecular level. It is responsible for the structural reorganisation of mixed molecules, and also determines the structure and properties of many biological molecules and macromolecules.Infrared spectroscopy plays a crucial role in the study of hydrogen bonding. To confirm the eutectic mixing of our active ingredient, we carried out an infra-red analysis using a Jasco FT/IR-4200 spectrophotometer, shown in (**Figure II.5**).

Figure II.5. Jasco FT/IR-420 spectrophotometer (Original, 2024).

II.4. Preparation of mother and daughter solutions of pure CXB and its five eutectic mixtures

II.4.1. Dissolution of the pure active principle and its eutectic mixtures with DMSO and preparation of stock solutions

Well-defined masses of the pure celecoxib and the five eutectic mixtures were weighed using an Ohaus analytical balance, accurate to 10^{-4}g and summarised **in (Table II.4).**

The weighed masses were dissolved in 10 ml of DMSO (Dimethylsulfoxide) and homogenised using a Vortex mixer.

Table II.4: Masses of pure active ingredient and eutectic mixtures

Component weighed	Mass (mg)	Concentration (mg/ml)
Pure celecoxib	903.8	90.38
Eutectic mixture 1	319	31.9
Eutectic mixture 2	461.3	46.13
Eutectic mixture 3	429.3	42.9
Eutectic mixture 4	566	56.6
Eutectic mixture 5	318	31.8

II.4.2. Preparation of daughter solutions by dilution of the stock solution to 1/2, 1/5 and 1/10

In order to determine the desired concentration range, we diluted the stock solution into three daughter solutions of concentrations 1/2, 1/5 and 1/10 respectively, as shown in **(Figure II.6)**. In addition, the positive control was diluted in a similar manner. The same volume (50µl) of DMSO (Dimethylsulfoxide) is used as the negative control.
The volumes we used are as follows:

- Volume of stock solution: 50 µl
- Daughter solution to 1\2: 2 ml eutectic mixture (mother solution)+ 2 ml DMSO.
- Daughter solution to 1:5: 1 ml crude eutectic mixture (stock solution)+ 4 ml DMSO.
- Daughter solution to 1:10: 1 ml eutectic mixture (stock solution)+ 9 ml DMSO.

Figure II.6. Dilution of the stock solution (Original, 2024).

II.5. Assessment antibacterial activity

Antibacterial activity was assessed using two methods:

- The solid-state diffusion method, which consists observing the zone inhibition of bacterial growth.

- The liquid dilution method, which is based on the determination of minimum inhibitory concentrations (MIC) and minimum bactericidal concentrations (MBC) of eutectic mixtures.

II.5.1. Method of distribution on disc

II.5.1.1. Preparation of the preculture

Antimicrobial tests need to be carried out on young bacterial cultures that are 18 to 24 hours old and in an exponential growth phase. The strains are regenerated by introducing the bacterial species into a culture medium (CM).

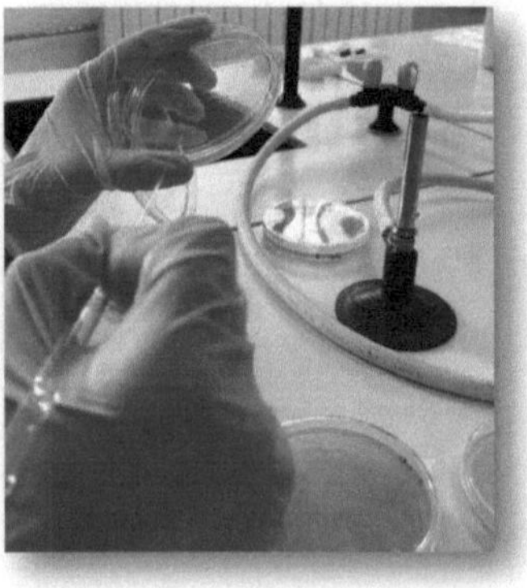

Figure II.7. Replication of bacterial strains (Original, 2024).

II.5.1.2. Preparation of the bacterial suspension

After growing the bacteria in MH agar medium, we selected 3 to 5 well-isolated, identical colonies. These were transferred to 5 ml of sterile physiological water and shaken using a vortex mixer. The bacterial suspension was standardised to a concentration of 10^6 CFU, corresponding to an optical density of 0.5 MFU (MCFarland unit), measured using a densitometer at a wavelength of 625 nm.

II.5.1.3. Seeding

Inoculation is carried out in petri dishes by distributing the suspension. bacterial over the entire surface of Miler Hinton (MH) culture medium, prepared according to the manufacturer's instructions. The seeded petri dishes are incubated at 37 °C for 24 hours for possible bacterial growth.

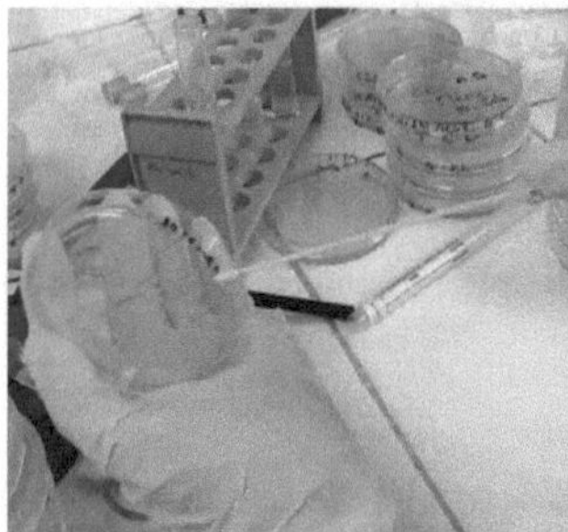

Figure II.8. Inoculation of bacteria (Original, 2024).

II.5.1.4. Preparing the discs

Once the seeding of bacteria in the MH culture medium has been established, Wattman No. 3 paper discs are placed on the surface of the agar, using sterilised forceps, and then soaked with 10 µl of crude and diluted eutectic mixture, metronidazole and DMSO as positive-negative controls respectively. Three soaked discs of each concentration were placed in the same petri dish. Petri dishes containing all concentrations of the different samples, are incubated at 37°C for 24 hours. Once incubation is complete, the inhibition diameters of the different bacteria tested are measured using a caliper.

Figure II.9. Used sterile discs (Original, 2024)

II.5.2. Liquid microdilution method

The microdilution method was used to determine the antibacterial efficacy of eutectic mixtures of celecoxib in a liquid medium. It was used to assess the minimum bactericidal concentrations (MBC) and minimum inhibitory concentrations (MIC) for the eutectic mixtures synthesised.

II.5.2.1. Determination of the minimum inhibitory concentration (MIC)

The MIC is the lowest concentration of antibiotic that inhibits visible growth after an incubation period of 18 to 24 hours. It is used to define the sensitivity or resistance of bacterial strains to antimicrobial agents **(Kablan et al, 2008)**.

II.5.2.2. Determining the minimum bactericidal concentration (MBC)

The minimal bactericidal concentration (MBC) is the concentration of an antimicrobial agent that leaves no more than 0.01 surviving germs. It is the lowest concentration necessary to kill the bacteria present in a culture.

II.5.2.3. Preparation of microplates

The MIC (Minimum Inhibitory Concentration) of the microorganisms studied was assessed in a sterile 96-well microplate, using a single plate for two eutectic mixtures. To do this, we proceeded as follows: The bacteria used in the microplate microdilution process are **Staphylococcus aureus** and **Escherichia coli**.

The microplates are prepared as follows:

Wells (A, B, C, F, G and H) **(Figure II.10)** are inoculated with 100 µl of liquid Mueller Hinton broth. To the 1st wells a volume of 100µl of eutectic mixture is added. A ½ dilution is performed on the five remaining wells. Wells D and E were left empty to avoid contamination. Wells 10 to 12 were kept for the positive control, in this case metronidazole, an antibacterial of the nitroimidazole family. Metronidazole was diluted 1/2 and 1/5 wells 11 and 12. Well 9 was dedicated to the negative control (DMSO). An incubation period of 18 hours at 37°C is essential reading the expected results.

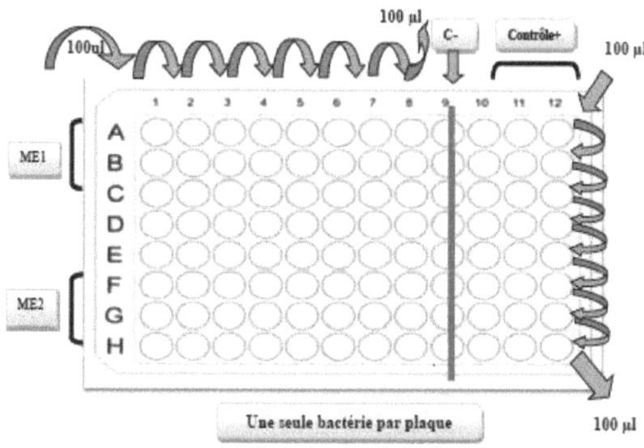

Figure II.10. Preparation of microplates

II.5.2.4. Evaluation of the CMB

Once incubation was complete, petri dishes seeded with bacteria and inoculated with a few microlitres of eutectic mixtures taken from the corresponding wells at concentrations equal to or greater than the MIC were incubated for 18 hours at 37°C to determine the BMC.

II.6. Determination of solubility

The gravimetric method was used to determine the solubility of pure and combined CXB in water at a temperature of 37°C. An excess of pure or mixed solid component was added to 4 ml of water, and the suspension was vortexed for a few minutes. The suspension was then placed in an incubator fitted with a shaker. The prepared solution was stirred for 24 hours to reach thermodynamic equilibrium. Once equilibrium was reached, the suspensions obtained were centrifuged at 3000 rpm for 30 min and filtered using a 0.45 µm syringe filter. The saturated solutions of CXB and eutectic CXB-coformer mixtures were kept in a fume hood for about a week to allow water to evaporate. The solubility of pure CXB (mg/ml) in water was calculated from the product of the mass of the dried material (solubility of the eutectic mixture) and the respective content of CXB in the dried eutectic mixture (in mass fraction).

RESULTS AND DISCUSSION

In this chapter, we present a discussion of the results obtained, covering the synthesis results, the characterisation, the evaluation of the antibacterial activity of the five eutectic mixtures synthesised and their solubility in water.

III.1. Synthesised eutectic mixtures

The binary mixtures prepared by wet grinding celecoxib with the various coformers, also known as inert carriers (salicylic acid, maleic acid, nicotinamide, succinic acid and cinnamic acid), chosen for their chemical structure, various functional groups and non-toxicity, are obtained in the form of a white powder (**Figure III.1**). This result can be translated into weak adhesive forces, due to hydrogen bonding, Van der Waals interactions, electrostatic forces and/or n-n interactions.

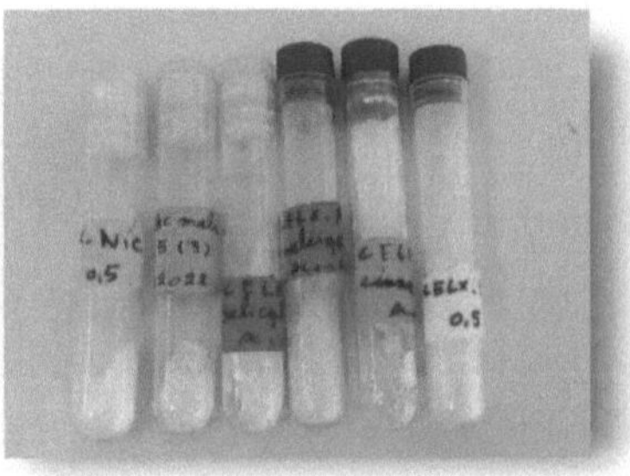

Figure III.1. physical and colour powders of the various eutectic mixtures synthesised (Original, 2024).

III.2. Characterisation

III.2.1. Differential Scanning Calorimetry (DSC)

The thermograms obtained by differential scanning calorimetry are shown in **Figure III.2**.

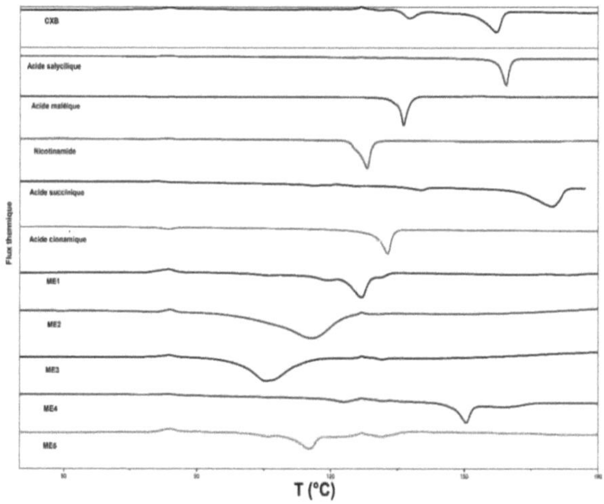

Figure III.2: Thermograms of the various pure substances and their eutectic mixtures

From the shape of the thermograms, we can see that the mixtures synthesised in a 1:1 stoichiometric ratio all exhibit a single thermal accident, characterised by melting temperatures lower than those of the pure substances taken separately. These results confirm the formation of eutectic mixtures. The melting temperatures of the different mixtures obtained experimentally are shown in **(Table III.1)**.

Table III.1: Melting temperatures of the pure substances and binary mixtures studied

Eutectic mixtures	Melting temperature (°C)
CXB-Maleic acid	124.8
CXB-Salicylic acid	100.44
CXB-Nicotinamide	96.02
CXB-Succinic acid	143.4
CXB-Cinnamic acid	111.8

III.2.2. XRD analysis

X-ray diffraction analyses play a crucial role in the study of eutectic mixtures, enabling the detection of changes in crystalline structures. **Figure III.3** shows the X-ray diffractograms of CXB, maleic acid and the eutectic mixture.

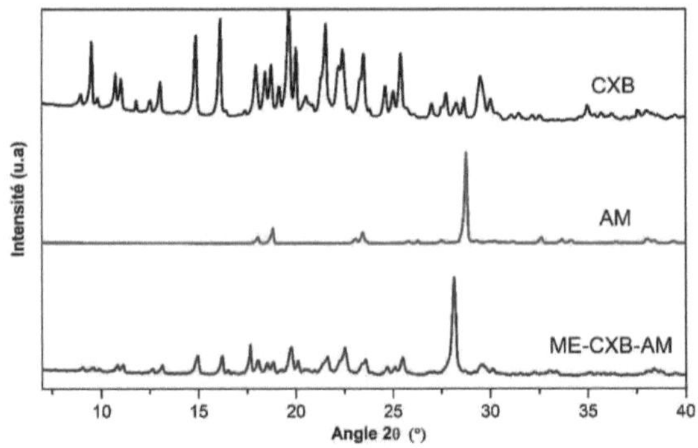

Figure III.3. Diffractograms of CXB, A.MAL and CXB-A.MAL

From Figure III.3, we can see that both the pure bodies and the resulting binary system are crystalline. The XRD analysis of the pure CXB revealed high intensity diffraction peaks at values of 2θ (15.1°, 16.2°, 19.6°, 20.1°, 21.6 and 22.4°). This is consistent with data reported in the literature **(M.Hawley et al ,2006 ; Z.Yang et al ,2022 ; G.Chawla et al ,2003).** Maleic acid shows distinct peaks at 2θ values of 9.8°, 21.6°, 22.8°, 25.3° and 29.5°. This result is in agreement with the literature **(AM.Araya-Sibaja et al ,2019)**. The eutectic mixture CXB-A.MAL retains the characteristic peaks of CXB and maleic acid without alterations in their position or shape. These findings confirm the absence of a new crystalline phase, resulting in the formation of a eutectic mixture and not a cocrystal.

III.2.3. Characterisation of synthesised eutectic mixtures by scanning electron microscopy

Figure III.4 shows some SEM micrographs of CXB, A.MAL and their mixture CXB-A.MAL. From the micrographs in Figure III.4, we can see that the eutectic mixture CXB-A.MAL has a different crystalline morphology to that of the pure bodies.

(a)

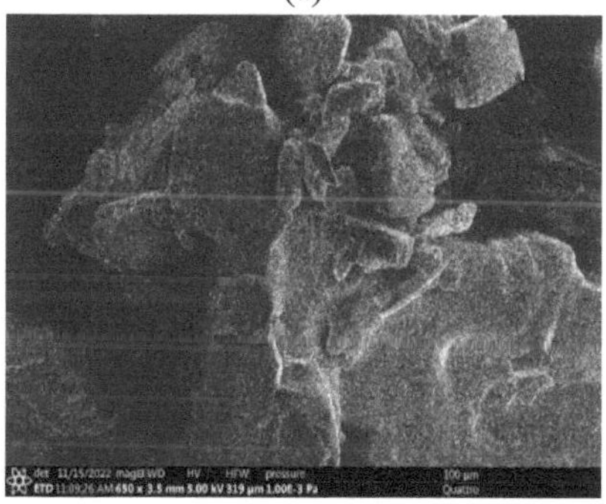

(b)

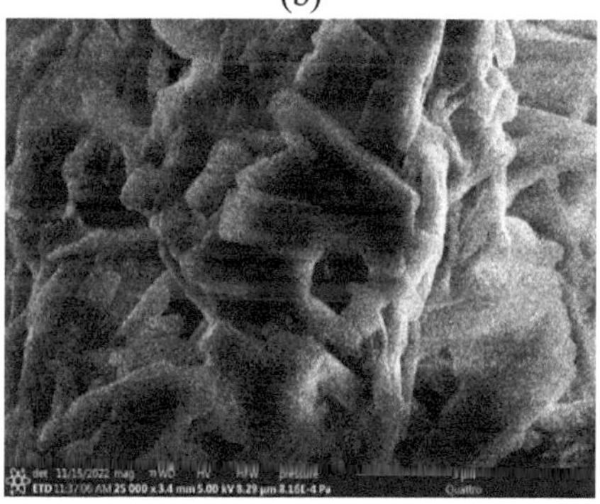

(c)

Figure III.4: Morphologies of (a) CXB, (b) A.MAL, (c) CXB-A.MAL.

III.2.4. Characterisation by FTIR

Infrared spectroscopy is an effective technique for intermolecular interactions in solids such as co-crystals and salts. It can be used to determine the nature of the interactions, by means of the different frequencies due to the vibration and elongation of the bonds established by the proton donor and acceptor functional groups.In order to analyse some of the cocrystals produced by wet grinding, the samples were finely ground and mixed with potassium bromide. The spectra were recorded with a resolution of 2 cm^{-1}. The infrared spectra of the eutectic mixture CXB-AM and their pure bodies are shown in **(Figure III.5)**.

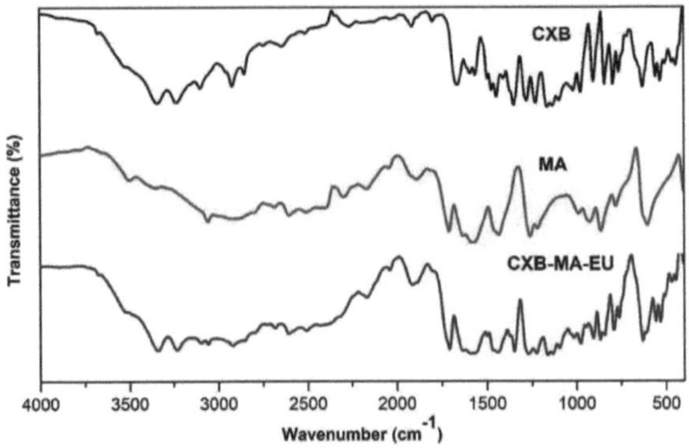

Figure III.5. Infrared spectra CXB, A.MAL and the binary mixture CXB-A.MAL

The FT-IR spectrum of pure CXB reveals peaks at 3230.6 cm^{-1} and 3339.78 cm^{-1} corresponding to symmetric and asymmetric elongation vibrations of the NH group respectively. An elongation vibration band of the -C=N imine group at a wave number of 1665.2 cm^{-1}. A band at 1344.3 cm^{-1} characteristic of a vibration by elongation of the -C=O bond of the sulphonamide group and a band at 785.41 cm^{-1}, corresponding to the aromatic -CH group. These spectral results are in line with those reported in the literature **(G.Lu et al ,2006)**. In the FT-IR spectrum of MA, peaks were observed at 3059.7 cm^{-1} corresponding to the vibration of the -OH group of the carboxylic acid, at 2879.83 cm^{-1} (vibration of the C-H bond), at 1704 cm^{-1} (C=O group), 1581 cm^{-1} (vibration by elongation of the C=C group of the cyclic alkene), 1341.79 cm^{-1} (vibration by deformation of the O-H group) and 1098.64 cm^{-1} (C-O), as reported in previous studies **(P.Tran et al,2023;V.M.Pandya et al ,2009)**.

The spectrum of the eutectic mixture revealed peaks corresponding to both CXB and maleic acid, with no peak shifts or broadening. This result can be explained by the absence of covalent interactions between the active ingredient and the coformer and confirms the presence of the original crystalline state of the pure compounds in the binary pharmaceutical system, without the formation of new molecules or alterations to intermolecular interactions.

III.3. Determination of the antibacterial activity of eutectic mixtures of celecoxib

III.3.1. Antibacterial activity in solid media

The antibacterial activity of the positive control (metronidazole), the different PA/inert carrier binary combinations at different concentrations and the DMSO used as a negative control, was evaluated on three bacterial strains**, namely Escherichia coli, Klebsiella pneumoniae** and **Staphylococcus aureus**.

III.3.1.1. Positive control results

Antibiograms in the presence of the positive control,this case pure metronidazole at a concentration of 60 mg/ml, were carried out on the three bacterial strains. The results obtained shown in (**Figure III.6**) all reveal the presence of a halo of inhibition of bacterial growth, corresponding to a diameter of inhibition of between 6.92 mm and 13.81 mm.
mm. The reactivity sequence of the positive control is as follows:

Klebsiella pneumoniae> Escherichia coli> Staphylococcus aureus

Staphylococcus aureus is the microorganism most resistant to pure metronidazole due to the low permeability of the antibacterial agent. These results are in line with the literature **(N. Islam et al, 2022).**

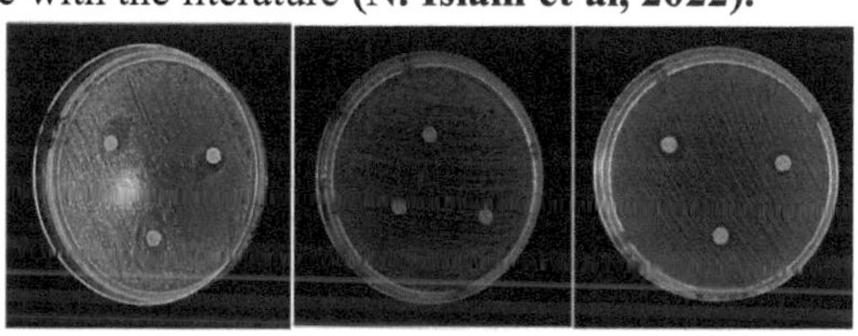

Klebsiella pneumoniae Staphylococcus aureusEscherichia coli

Figure III.6: Effect of the pure MTZ positive control on the three bacterial strains tested.

III.3.1.2. Negative control results

Figure III.7 shows that no growth inhibition was observed for the negative control. DMSO had no effect on the growth of any bacteria on agar. This confirms the correct choice of solvent used for solubilisation.

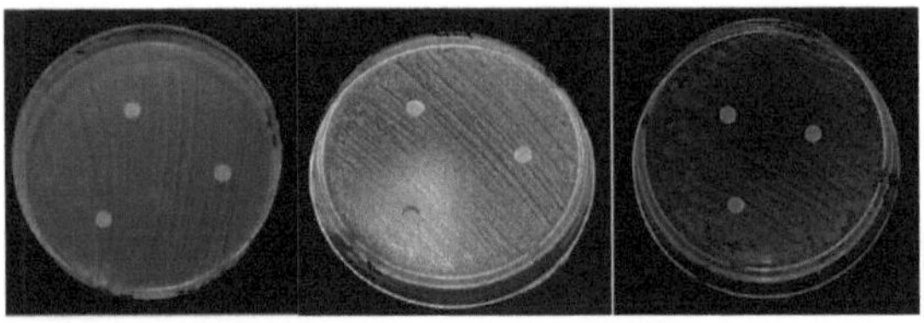

Klebsiella pneumoniae Escherichia coli Staphylococcus aureus
Figure III.7: Effect of the negative control (DMSO) on the four bacterial strains tested.

III.3.1.3. Antibacterial activity of different eutectic mixtures on discs

The antibiograms of the various crude and diluted binary mixtures studied and formed from the PA/coformer binary combination are summarised in (**Table III.2)** and illustrated in (**Figure III.8**).

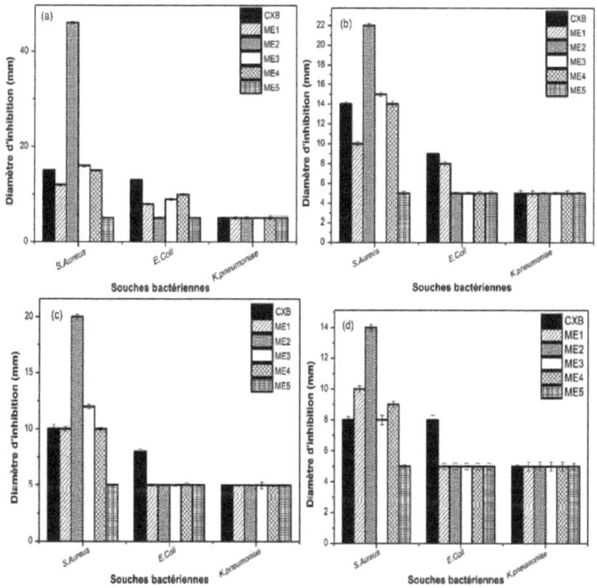

Figure III.8. Histograms of the antibiogram results for the five eutectic mixtures in relation to the bacterial strains tested: **(a)** Comparative histogram of the five binary mixtures with a concentration of C_0; **(b)** Comparative histogram of the five diluted binary mixtures with a concentration of $C_{(0)}/2$; **(c)** Comparative histogram of the five diluted binary mixtures with a concentration of $C_0/5$; **(d)** Comparative histogram of the five diluted binary mixtures with a concentration of $C_0/10$.

From the results obtained we can deduce the following:

❖ Raw celecoxib acts on the bacterial strains tested in this study in the following sequence:

Staphylococcus aureus> Escherichia coli >Klebsiella pneumoniae

These results are in agreement with the work of **(M.Hayder et al.,2013)**. However, the mechanism

The antibacterial action of celecoxib is poorly understood. Several hypotheses have been put forward:

- Interaction with bacterial cell membrane lipids, disrupting their integrity and function;
- Inhibition of prostaglandin synthesis, which could have an indirect effect on

bacterial growth;

• Disruption of bacterial cell signalling pathways, affecting their metabolism and survival;

• Modulation of the expression of certain bacterial genes, thus generating its activity antibacterial.

Inflammation promotes bacterial growth by causing an accumulation of fluid in the area of injurydue to increased vascular permeability, leading to oedema, which can actually promote bacterial growth and cause tissue damage, providing a good environment and nutrients for bacteria. As a result, CXB inhibits bacterial growth by inhibiting the inflammatory process. In addition, COX-2 deficiency leads to a reduction in inflammation and cytokine release.

❖ The eutectic mixtures **CXB-A.MAL**, **CXB-A.SAL**, **CXB-NIC** and **CXB-A.SUC** show the following characteristics inhibition diameters ranging from 8 to 46 mm. The above non-antibacterial agents have significant antibacterial activity against **S.Aureus** according to the sequence of reactivity:

CXB-A.MAL> CXB-NIC> CXB-A.SAL> CXB-A.SUC

This reactivity can be reflected in the chemical structure and polarity of these mixtures, as well as the nature of the intermolecular interactions at play in these binary combinations, and in particular hydrophobic interactions, which increase antibacterial activity. The highest antibacterial activity was obtained in the presence of ME2 (CXB-A.MAL), with an inhibition diameter of 46 mm recorded against S. Aureus. This high inhibition halo can be explained by the stability of ME2, due to the establishment of various non-covalent interactions, namely Van der Waals interactions, n-n interactions and hydrogen bonding between the proton donor and acceptor groups the two molecules.

S. Aureus, a gram-positive bacterium, is characterised by a thick peptidoglycan wall, which facilitates the insertion and solubilisation of the hydrophobic mixture in the plasma membrane, resulting in the loss of the entire lipid bilayer and a change in its permeability to ions and other cellular constituents, leading to the loss of cytoplasmic elements, lysis and cell death. Also, disruption of the wall structure or inhibition of specific enzymes involved in peptidoglycan biosynthesis allows the antibacterial agent to reach the micro-organism's DNA. Once there, the antibacterial agent binds to the nucleic proteins and inhibits the synthesis proteinsresulting in the inhibition or death of the micro-organism.

❖ ME5 synthesised by the combination of CXB and cinnamic acid has no effect. antibacterial effect on **S.Aureus**. This null effect can be translated by a high polarity and consequently a low hydrophobicity, which reduces its interaction with the polysaccharide chains making up the wall, thus creating impermeability to the ME5.

❖ The effect of the eutectic mixtures synthesised on gram-negative bacterial strains is as follows

less pronounced than the effect obtained against **S.Aureus**. This result is in line with the literature **(D. Mishra et al ,2010)**.

❖ The gram-negative bacteria tested in this work, namely E.Coli and K.pneumoneae are

respectively less sensitive to resistant. Pure CXB and the eutectic mixtures synthesised show inhibition diameters ranging from 8 to 13 mm against **Escherichia coli**. This low sensitivity may be due to the nature of the cell wall, whose unique characteristics often make gram-negative bacteria more resistant to many hydrophilic agents. Nevertheless, the antibacterial effect observed may be due to the action of the agent with the lipids of the cell membrane, resulting in its disruption and consequently becoming permeable, causing leakage of essential cellular components such as ions, nucleotides and metabolites.

❖ Membrane disruption can also affect critical membrane functions,

such as nutrient transport and waste elimination, leading increased stress and cell death.

❖ **Klebsiella pneumoniae** is a strain resistant to pure CXB and its eutectic mixtures with zero halo inhibition due to target modification or inactivation enzyme. This result is in line with the literature **(M.Hayder et al ,2013).**

Table III.2: Results of antibacterial activity tests obtained by diffusion in solid medium of the five agents synthesised.

Inhibition zone diameter in (mm)																								
	Pure celecoxib				CXB+A.SAL				CXB+A.MAL				CXB+NIC				CXB+A.SUCC				CXB+A.CIN			
	Mother solution	1/2	1/5	1/10	Stock solution	1/2	1/5	1/10	Stock solution	1/2	1/5	1/10	Mother solution	1/2	1/5	1/10	Stock solution	1/2	1/5	1/10	Stock solution	1/2	1/5	1/10
E. coli	13	9	9	8	8	8	5	5	5	5	5	5	9	5	5	5	10	5	5	5	5	5	5	5
S. aureus	15	14	10	8	12	10	10	10	46	22	20	14	16	15	12	8	15	14	10	9	5	5	5	5
K pneumoniae	5	5	5	5	5	5	5	5	5	5	5	5	5	5	5	5	5	5	5	5	5	5	5	5

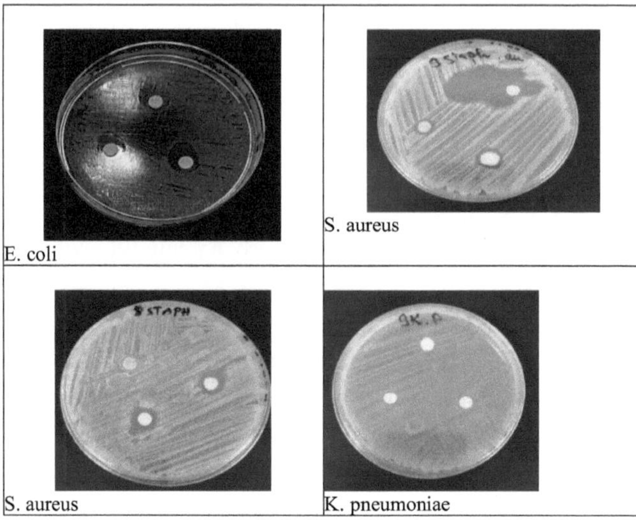

Figure III.9. Antibacterial effect of ME2 on the three strains tested.

III.3.2. Determination of bactericidal minimum inhibitory concentrations (MICs)

The MIC is defined as the lowest concentration of inhibition of bacterial growth, and is assessed by the microdilution method on microplates. This method revealed some transparent wells and others that were cloudy, due to an insufficient concentration of the synthesised compound required to inhibit bacterial growth. **Figure III.10** illustrates the histograms corresponding to the MIC obtained for pure CXB and its eutectic mixtures. The results obtained for MIC and BMC are summarised in (**Table III.3**) and represented in (**Figure III.11-III.13**).

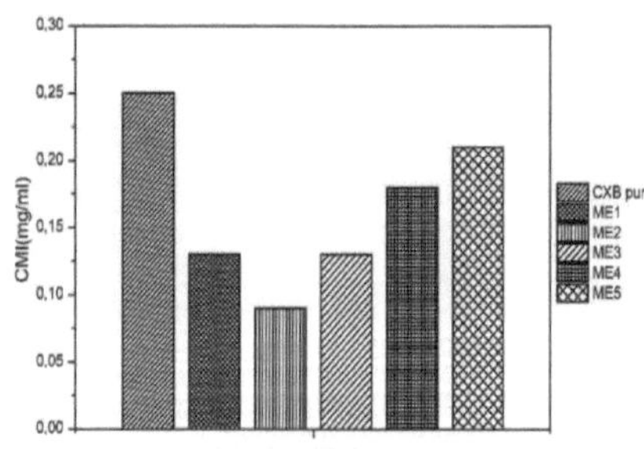

Figure III.10. MIC of pure CXB and its five eutectic mixtures

Table III.3 MIC (mg/ml) and BMC (mg/ml) obtained by the microdilution method for the two bacterial strains

Constituent	Microbial strains	MIC(mg/ml)	CMB
Pure CXB	E. coli	0.028	>0.028
	S. aureus	0.028	>0.028
CXB+ A.SAL	E. coli	0.0099	>0.0099
	S. aureus	0.0099	>0.0099
CXB+ A.MAL	E. coli	0.0144	>0.0144
	S. aureus	0.0144	> 0.0144
CXB+ NIC	E. coli	0.0134	> 0.0134
	S. aureus	0.0134	> 0.0134
CXB+ A.SUC	E. coli	0.0353	> 0.0353
	S. aureus	0.0353	> 0.0353
CXB+ A.CIN	S. aureus	0.0198	> 0.0198

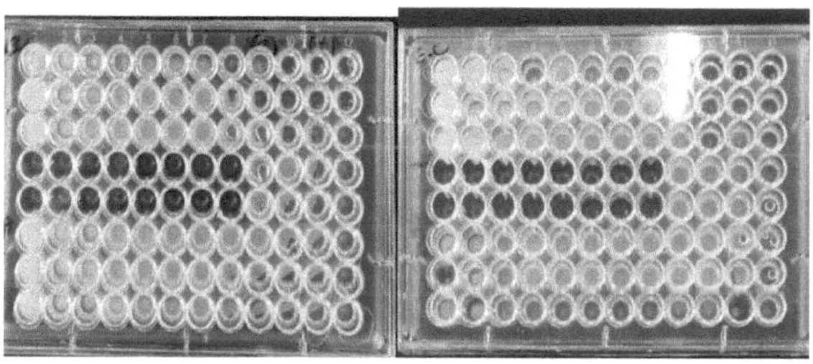

(a) (b)

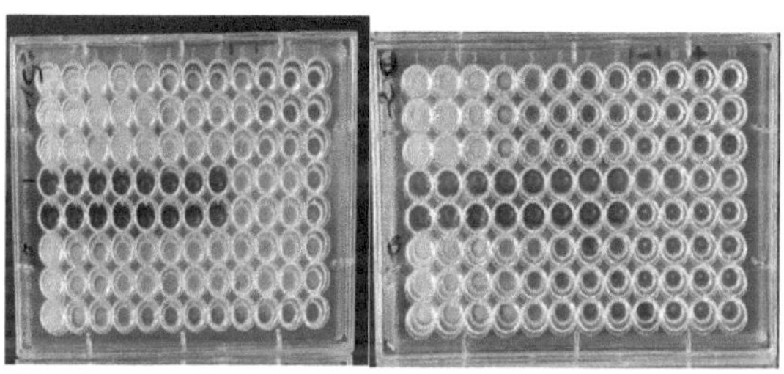

(b) (d)

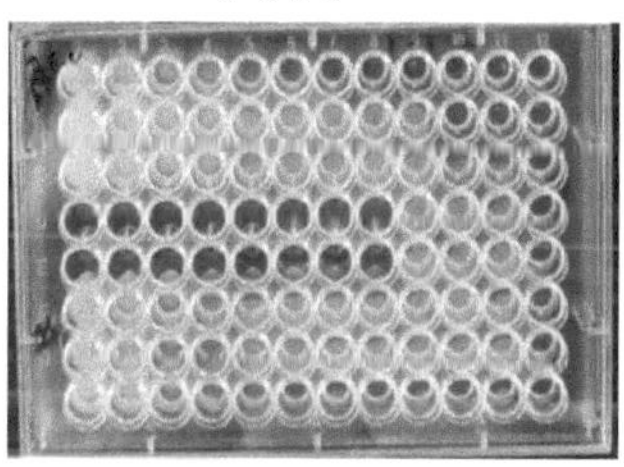

(e)

Figure III.11. Determination of MIC and MBC of the five eutectic mixtures studied against: (a,b) E.Coli and (c,d,e) S.Aureus

According to the literature **(N. Djabou et al, 2013)** the bacteria are :

□ Resistant when their MIC is between 25 and 50 mg/ml.

□ Moderately sensitive (+) for a value between 3 and 12.5 mg/ml. □ Sensitive (++) for a MIC between 0.4 and 2 mg/ml.

□ Extremely sensitive when their MIC less than or equal to 0.2 mg/ml.

The minimum inhibition concentrations evaluated in this study varied between 0.498 and 1.768 mg/ml. These values reflect the sensitivity of the bacteria tested via the eutectic mixtures synthesised, which can be characterised bacteriostatic and bactericidal agents at higher concentrations.

A MIC of around 0.49 mg/ml was obtained with the binary combination CXB-A.MAL against bacterial strains (E. Coli and Staphylococcus aureus). This result is in agreement with the inhibition haloes obtained by the solid-state diffusion method. The sulphonamide group of celecoxib can interact via hydrogen bonding with the carboxyl and hydroxyl groups of maleic acid. These interactions create a stable mixture, helping to reduce the melting point and therefore increase therapeutic efficacy.
The highest MICs were recorded in the eutectic mixtures CXB-A.CIN and CXB- A.SUC.
The MBC makes it possible to deduce the nature of the agent used for any antibacterial effect. If this concentration is equal to the MIC, the agent is bactericidal.
If the MBC is greater than the MIC, it is bacteriostatic. We can deduce from our results that, at the concentrations taken, our agents behave as bacteriostatic.

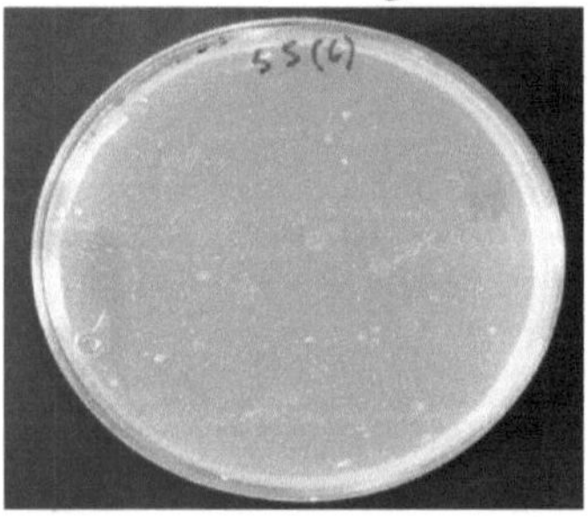

Figure III.12. CMB of ME 2 against the bacterial strain (Staphylococcus aureus).

Figure III.13. CMB of different bacterial strains

III.4. Determination of solubility

Solubility is defined as the maximum quantity of a substance that will dissolve in a given quantity of solvent at a given temperature and pressure. It is a characteristic property of a specific solute-solvent combination **(B.Mittal et al.,2016).** Thermodynamic solubility represents the concentration of a compound in equilibrium with an excess of undissolved solid phase. The test is typically performed by adding an excess amount of solid to the solubility medium and mixing for extended periods, typically 24-48 hours, until equilibrium is achieved.

An active substance is considered to have a high solubility when the maximum administrable dose by immediate release is soluble in 250mL of an aqueous medium with a pH ranging from 1 to 6.8 at 37±1°C. When more than 85% of the dose of active substance initially administered is absorbed through the intestinal barrier, it has high permeability.

Gravimetrically determined water solubility results for the synthesised pure compounds and eutectic mixtures are summarised **in (Table III.4)** and shown in **(Figure III.14).**

Table III.4. Solubility results for the various pure bodies and eutectic systems synthesised.

Constituents	Solubility (mg/ml)
Celecoxib (CXB)	6.8765
Maleic acid (A.MAL)	15
Salicylic acid (A.SAL)	0,0466
Nicotinamide (NIC)	33
Cinnamic acid (A.CIN)	2,37
CXB- A.MAL	13,48
CXB- A.SAL	5,186
CXB- NIC	15,89
CXB- A.CIN	5,84

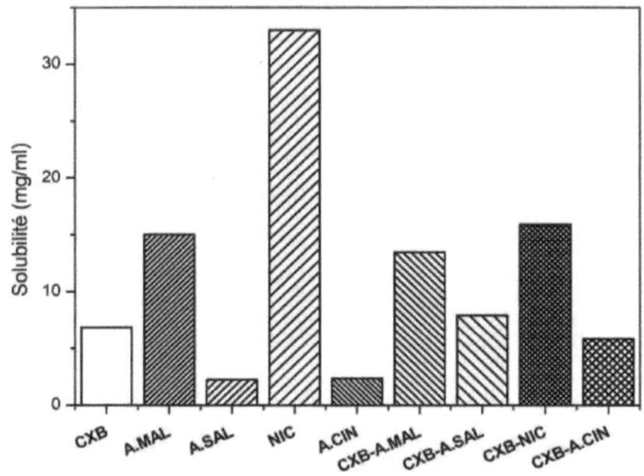

Figure III.14. Solubilities of the pure compounds and eutectic mixtures synthesised.

The solubility of CXB increased markedly in the eutectic mixtures: CXB-A.MAL, CXB-NIC. As a result, it increased by a factor of about 2 for the CXB-A.MAL system and by a factor of about 2.5 for the CXB-NIC system. These improvements can be explained by the following mechanisms:

- Use a highly water-soluble coformer such as nicotinamide ;
- Reduced particle size of the active ingredient and improved wettability of its crystals in the medium;
- Impact of the hydrophilic nature of the coformer on improving solubility.

CONCLUSION

This work is part of a PRFU research project focusing on the synthesis, characterisation, improved solubility and assessment of the biological activities of cocrystals and eutectic mixtures of active pharmaceutical ingredients, produced by a binary PA/coformer combination with different chemical structures and physicochemical properties, leading to heterosyntons with significantly improved solubility, bioavailability and therapeutic effect. The strategy used in this study consists of repositioning a non-antibacterial active ingredient, celecoxib (CXB), used in a binary combination with five coformers, namely salicylic acid, maleic acid, nicotinamide, succinic acid and cinnamic acid, to combat the antibiotic resistance of certain multi-resistant bacterial strains and improve the solubility of CXB. The binary combinations CXB-A.SAL, CXB-A.MAL, CXB-NIC, CXB-A.SUC and CXB-A.CIN were synthesised by solvent assisted grinding in a 1:1 stoichiometric ratio and characterised by DSC, XRD, SEM and FTIR.The antibacterial activity of pure CXB and its eutectic mixtures developed using the bacterial strains Escherichia coli, Klebsiella pneumoniae and Staphylococcus aureus was assessed using two methods: the solid-state diffusion method and microdilution in microplates, to determine the diameter of the zones of inhibition and the minimum inhibitory and bactericidal concentrations respectively. The thermograms of the five binary combinations synthesised in this work, obtained by DSC, showed the appearance of a single thermal accident, characterised by a melting temperature lower than that of the pure bodies used in the preparation. These results can be explained by the formation of eutectic mixtures via non-covalent interactions, in this case hydrogen bonding, n-n interactions, electrostatic forces and Van der Waals interactions. XRD and FTIR analyses enabled us to confirm the formation of eutectic mixtures. The diffractograms obtained by XRD and the FTIR spectra revealed, respectively, the absence of a new crystalline phase and the existence of the characteristic bands of CXB and the coformer without alteration or broadening. The antibacterial activity of the eutectic mixtures, assessed in vitro by the disk diffusion method, against the three bacterial strains tested, follows the reactivity sequence:

Staphylococcus aureus> Escherichia coli> Klebsiella pneumoniae

The non-antibacterial agents **CXB-A.MAL**, **CXB-A.SAL**, **CXB-NIC** and **CXB-A.SUC** have the following properties inhibition diameters ranging from 8 to 46 mm, in the following sequence:

CXB-A.MAL> CXB-NIC> CXB-A.SAL> CXB-A.SUC

This reactivity can be reflected in the chemical structure and polarity of these mixtures, as well as in the nature of the intermolecular interactions at play in these binary combinations, and in particular the hydrophobic interactions that increase the antibacterial activity. The highest antibacterial powers were obtained in the presence of ME2 (CXB- A.MAL), with an inhibition diameter of 46 mm, recorded against S. Aureus, explained by the stability of the ME2 heterosyntons. Gram-negative bacterial strains are less sensitive to pure CXB and its eutectic mixtures, with inhibition zone diameters of between 8 and 13 mm. These results are in line with the literature. The MIC values obtained ranged from 0.09 to 0.72 mg/ml. This variation allowed us to deduce that the eutectic mixtures synthesised are bacteriostatic at the minimum inhibitory concentration and bactericidal at a concentration above the MIC. In order to improve the bioavailability and, consequently, the therapeutic effect of CXB, we determined its solubility in water using the gravimetric method. Water solubility increased significantly in the eutectic mixtures: CXB-A.MAL, CXB-NIC, due the reduced particle size of the active ingredient and the use of hydrophilic coformers.

Looking ahead

Further studies could be envisaged to explore in greater depth some of the points raised this manuscript.

It would therefore be interesting to :

- Repositioning a variety of non-antibacterial agents in the fight against bacteria multidrug-resistant ;
- Developing eutectic mixtures with the formation of heterosyntons with different physico-chemical and therapeutic properties;
- Evaluate their solubility in a range of deep eutectic solvents (DES).

REFERENCES

A

Al-Kuraishy, H. M, Algareeb, A, Al-windy, A.S, Experimental antibacterial activity of selective cyclooxygenase antagonist, Int J Basic Clin Pharmacol. 2013 Aug;2(4), 381-385

Alshaikh, R.A., Essa, E.A., El Maghraby, G.M., 2019. Eutexia for enhanced dissolution rate and anti-inflammatory activity of nonsteroidal anti-inflammatory agents: Caffeine as a melting point modulator. Int. J. Pharm. 563, 395-405. https://doi.org/10.1016/j

Amanda Koberstain Surur Morandin Ferrisse , Analú Barros de Oliveira , Carla Raquel Fontana, Bacterial resistance to antimicrobial photodynamic therapy: A critical update. Journal of Photochemistry & Photobiology, B: Biology 255 (2024) 112905

Araya-Sibaja, A., Vega-Baudrit, J., Guillén-Girón, T., Navarro-Hoyos, M., Cuffini, S., 2019. Drug solubility enhancement through the preparation of multicomponent organic materials: eutectics of lovastatin with carboxylic acids. Pharmaceutics 11, 112. https://doi.org/10.3390/pharmaceutics11030112.

B

Blain, H., Jouzeau, J. Y., Netter, P., & Jeandel, C. (2000). Les anti-inflammatoires non stéroïdiens inhibiteurs sélectifs de la cyclooxygénase 2. Intérêt et perspectives. La revue de médecine interne, 21(11), 978-988.

Bouskraoui M, Zouhair S, Soora N, Benaouda A, Zerouali K, Mahmoud M. A practical guide to pathogenic bacteria. 2017.

Bursaux, E. (1994). Understanding cyclooxygenases to better control the effects of non-steroidal anti-inflammatory drugs.

C

Cervello, M., Bachvarov, D., Cusimano, A., Sardina, F., Azzolina, A., Lampiasi, N., ... & Montalto, G. (2011). COX-2-dependent and COX-2-independent mode of action of celecoxib in human liver cancer cells. Omics: a journal of integrative biology, 15(6), 383-392.

Chadha, K., Karan, M., Chadha, R., Bhalla, Y., Vasisht, K., 2017a. Is failure of cocrystallization actually a failure; Eutectic formation in cocrystal screening of hesperetin. J. Pharm. Sci. 106, 2026-2036. https://doi.org/10.1016/j.xphs.2017.04.038

Chadha, R., Sharma, M., Haneef, J., 2017b. Multicomponent solid forms of felodipine: preparation, characterisation, physicochemical and in-vivo studies. J. Pharm. Pharmacol. 69, 254-264. https://doi.org/10.1111/jphp.12685.

Chiu, H. C., Lee, S. L., Kapuriya, N., Wang, D., Chen, Y. R., Yu, S. L., ... & Chen, C. S. (2012). Development of novel antibacterial agents against methicillin-resistant Staphylococcus aureus. Bioorganic & medicinal chemistry, 20(15), 4653-4660

Chiu, H. C., Yang, J., Soni, S., Kulp, S. K., Gunn, J. S., Schlesinger, L. S., & Chen, C. S. (2009). Pharmacological exploitation of an off-target antibacterial effect of the cyclooxygenase-2 inhibitor celecoxib against Francisella tularensis. Antimicrobial agents and chemotherapy, 53(7), 2998-3002.

D

D'Costa V, King C, Kalan L, Morar M, Sung W, Schwarz C, et al. Antibiotic resistance is ancient. Nature. 2011; 477:457-61.

Dadiboyena, S., & T Hamme II, A. (2012). Synthesis of celecoxib and structural analogs-a review. Current Organic Chemistry, 16(11), 1390-1407.

DB Salunke, BG Hazra, VS Pore (2006) Steroidal conjugates and their pharmacological applications, Current medicinal chemistry 13 (7), 813-847

D.Muster, Anti-inflammatory , Volume 1, Issue 1, March 2005, Pages 21-29

E

Emami, S., Siahi-Shadbad, M., Barzegar-Jalali, M., Adibkia, K., 2018. Characterizing eutectic mixtures of gliclazide with succinic acid prepared by electrospray deposition and liquid assisted grinding methods. J. Drug Deliv. Sci. Technol. 45, 101-109. https://doi.org/10.1016/j.jddst.2018.03.006.

F

Fares E.M. Ali, Mustafa Ahmed Abdel-Reheim , Emad H.M. Hassanein ,Mostafa K.Abd El- Aziz, Hanan S. Althagafy, Khalid S.A. Badran. Exploring the potential of drug repurposing for liver diseases: A comprehensive study; Life Sciences 347 (2024) 122642.

G

Giovana Carolina Bazzo, Bianca Ramos Pezzini, Hellen Karine Stulzer, Eutectic mixtures as an approach to enhance solubility, dissolution rate and oral bioavailability of poorly water-soluble drugs; International Journal of Pharmaceutics 588 (2020) 119741

Goud, N.R., Suresh, K., Sanphui, P., Nangia, A., 2012. Fast dissolving eutectic compositions of curcumin. Int. J. Pharm. 439, 63-72. https://doi.org/10.1016/j.ijpharm. 2012.09.045.

H

Haneef, J., Chadha, R., 2017. Drug-drug multicomponent solid forms: cocrystal, coamorphous and eutectic of three poorly soluble antihypertensive drugs using mechanochemical approach. AAP PharmSciTech 18, 2279-2290. https://doi.org/10.1208/s12249-016-0701-1

Haneef, J., Chadha, R., 2018. Antioxidant-based eutectics of irbesartan: viable multicomponent forms for the management of hypertension. AAPS PharmSciTech 19,1191-1204. https://doi.org/10.1208/s12249-017-0930-y

Hao-Chieh Chiu,1 Jian Yang,1 Shilpa Soni,2,3 Samuel K. Kulp,1 John S. Gunn,2,3,4 Larry S. Schlesinger, and Ching-Shih Chen1 Pharmacological Exploitation of an Off-Target Antibacterial Effect of the Cyclooxygenase-2 Inhibitor Celecoxib against Francisella tularensis_ ANTIMICROBIAL AGENTS AND CHEMOTHERAPY, July 2009, p. 2998-3002 Vol. 53, No. 7

H Blain, J.Y Jouzeau, P Netter , C Jeandel Les anti-inflammatoires non stéroïdiens inhibiteurs sélectifs de la cyclooxygénase 2. Interest and future prospects Non-steroidal anti-inflammatory drugs with selective inhibitory activity on cyclooxygenase 2. Interest and future prospects.

Hyun, S.-M., Lee, B.J., Abuzar, S.M., Lee, S., Joo, Y., Hong, S.-H., Kang, H., Kwon, K.-A., Velaga, S., Hwang, S.-J., 2019. Preparation, characterization, and evaluation of celecoxib eutectic mixtures with adipic acid/saccharin for improvement of wettability and dissolution rate. Int. J. Pharm. 554, 61-71. https://doi.org/10.1016/J.IJPHARM.2018.10.044

K

Kablan B, Adiko M, Abrogoua D. In vitro evaluation of the antimicrobial activity of Kalanchoe crenata and Manotes longiflora used in ophthalmia in Côte d'Ivoire. 2008;6(5):282-8.

L

La Revue de Médecine Interne Volume 21, Issue 11, November 2000, Pages 978-98

Levoin, N. (2002). Reactive metabolites of non-steroidal anti-inflammatory drugs: Bases structurales de leurs interactions avec les cibles protéiques impliquées dans les processus inflammatoires (Doctoral dissertation, Université Henri Poincaré-Nancy 1).

Lode H. Clinical impact of antibiotic-resistant Gram-positive pathogens. Clin Microbiol Infect. 2009 ;15 :212-7.

M

Márió Gajdács, Non-antibiotic compounds affecting the growth of urinary pathogens during urine culture: a preliminary in vitro study, APH 2020;90:185-191

Malaise, M. (1996). Les anti-inflammatoires non stéroïdiens. Revue Médicale de Liège, 51(1).

Maragakis L, Perencevich E, Cosgrove S. Clinical and economic burden of antimicrobial resistance. Expert Rev Anti Infect Ther. 2008; 6:751-63.

Mishra D, Sudhir P, Panda P.K. EVALUATION OF ANTIMICROBIAL ACTIVITY OF SELECTIVE COX-2 INHIBITOR. Rasayan J.Chem, Vol.3, No.3 (2010), 532-538

N

Narva Deshwar Kushwaha, Joshua Nwabuife, Jivanka Mohan, Neil Koorbanally, Babita Kushwaha, Anil A. Chuturgoon, Terisha Ghazi A comprehensive review on the global efforts on vaccines and repurposed drugs for combating COVID-19 European Journal of Medicinal Chemistry 260 (2023) 115719

P

Pillon, F. (2014). Non-steroidal anti-inflammatory drugs. Actualités Pharmaceutiques, 53(534), 43-46.

R

Riekes, M.K., Engelen, A., Appeltans, B., Rombaut, P., Stulzer, H.K., Van den Mooter, G., 2016. New perspectives for fixed dose combinations of poorly water-soluble compounds: a case study with ezetimibe and lovastatin. Pharm. Res. 33, 1259-1275. https://doi.org/10.1007/s11095- 016-1870-z.

RN Saha, C. Sajeev, PR. Jadhav, SP Patil, N. Srinivasan (2002). Determination of celecoxib in pharmaceutical formulations using UV spectrophotometry and liquid chromatography.

S

Said Moshawih, Qais Jarrar, Abdul Alim Bahrin, Ai Fern Lim, Long Ming, Hui Poh Goh, Evaluating NSAIDs in SARS-CoV-2: Immunomodulatory mechanisms and future therapeutic strategies, Heliyon 10 (2024) e25734

Sang-Min Hyun, Benjamin Joon Lee, Sharif Md Abuzar, Soohun Lee, Yechan Joo, Seung- Hyeon Hong, Han Kang, Kyung-Ae Kwon, Sitaram Velaga, Sung-Joo Hwang, Preparation, characterization, and evaluation of celecoxib eutectic mixtures with adipic acid/saccharin for improvement of wettability and dissolution rate, https://doi.org/10.1016/j.ijpharm.2018.10.044

S. Alghadeer, K. Aljuaydi, S. Babelghaith, A. Alhammad, M. Alarifi, Self-medication with antibiotics in Saudi Arabia, Saudi Pharm. J. 26 (2018), https://doi.org/10.1016/j.jsps.2018.02.018

Shankar Thangamani, Waleed Younis and Mohamed N. Seleem (2015), Repurposing celecoxib as a topical antimicrobial agent, Front. Microbiol. 6:750. doi: 10.3389/fmicb.2015.00750

Srinivasu, M. K., Rao, D. S., & Reddy, G. O. (2002). Determination of celecoxib, a COX-2 inhibitor, in pharmaceutical dosage forms by MEKC. Journal of pharmaceutical and biomedical analysis, 28(3-4), 493-500.

SZ Alshawwa, AA Kassem, RM Farid, SK Mostafa, GS Labib (2022).

Nanocarrier drug delivery systems: characterization, limitations, future perspectives and implementation of artificial intelligence Pharmaceutics 14 (4), 883.

T

Thayyil AR, Juturu T, Nayak S, Kamath S. Pharmaceutical co-crystallization: Regulator aspects, design, characterization, and applications. 2020 ;10(2) :203.

Tive, L. (2000). Celecoxib clinical profile. Rheumatology, 39(suppl_2), 21-28.
Tołoczko-Iwaniuk, N., Dziemiańczyk-Pakieła, D., Nowaszewska, B. K., Celińska-Janowicz, K., & Miltyk, W. (2019). Celecoxib in cancer therapy and prevention-review. Current Drug Targets, 20(3), 302-315.

Z

Zhengge Yang, Ru Ma, Yin Chen, Yifang Zhang, Xin Liu, Bi-Feng Liu, Guisen Zhang, Chao Hao; Drug-drug eutectic mixtures of celecoxib with tapentadol and milnacipran which could improve analgesic and antidepressant efficacy, Journal of Drug Delivery Science and 67 (2022) 10299.

MIX
Papier aus verantwortungsvollen Quellen
Paper from responsible sources
FSC® C105338

Printed by Books on Demand GmbH, Norderstedt / Germany